SPOTLIGHT ON GLOBAL ISSUES

CLEAN AIR FOR ALL

Kathy Furgang

ROSEN PUBLISHING
NEW YORK

Published in 2022 by The Rosen Publishing Group, Inc.
29 East 21st Street, New York, NY 10010

Editor: Theresa Emminizer
Book Design: Michael Flynn

Photo Credits: Cover Gino Santa Maria/Shutterstock.com; (series globe background) photastic/Shutterstock.com; p. 4 kondrytskyi/Shutterstock.com; p. 5 leungchopan/Shutterstock.com; p. 6 photo one/Shutterstock.com; p. 7 Piupiupics/Shutterstock.com; p. 9 courtesy of Library of Congress; p. 10 Toa55/Shutterstock.com; p. 12 Alfred Eisenstaedt/The LIFE Picture Collection/Getty Images; p. 13 Mirrorpix/Getty Images; p. 14 https://en.wikipedia.org/wiki/Harry_S._Truman#/media/File:TRUMAN_58-766-06_(cropped).jpg; p. 15 https://en.wikipedia.org/wiki/Clean_Air_Act_(United_States)#/media/File:Clean_Air_Act_Signing.jpg; p. 16 Martin Good/Shutterstock.com; p. 17 Harold Cunningham/Getty Images; p. 18 Tanakorn Akkarakulchai/Shutterstock.com; p. 19 ivan_kislitsin/Shutterstock.com; p. 20 Denis Burdin/Shutterstock.com; p. 21 katz/Shutterstock.com; p. 23 Pablo Blazquez Dominguez/Getty Images; p. 24 Killua X/Shutterstock.com; p. 26 lev radin/Shutterstock.com; p. 27 picture alliance/Getty Images; p. 28 Mike Jett/Shutterstock.com; p. 29 Daniele COSSU/Shutterstock.com.

Library of Congress Cataloging-in-Publication Data

Names: Furgang, Kathy, author.
Title: Clean air for all / Kathy Furgang.
Description: New York : Rosen Publishing, [2022] | Series: Spotlight on global issues | Includes bibliographical references and index.
Identifiers: LCCN 2019051821 | ISBN 9781725323025 (library binding) | ISBN 9781725322998 (paperback) | ISBN 9781725323001 (6 pack)
Subjects: LCSH: Air--Pollution--Juvenile literature. | Air quality management--Juvenile literature.
Classification: LCC TD883.13 .F87 2022 | DDC 363.739/2--dc23
LC record available at https://lccn.loc.gov/2019051821

Manufactured in the United States of America

Some of the images in this book illustrate individuals who are models. The depictions do not imply actual situations or events.

CPSIA Compliance Information: Batch #CSR22. For further information contact Rosen Publishing, New York, New York at 1-800-237-9932.

CONTENTS

CHAPTER ONE

BREATHE IN

Breathe in and then slowly breathe out. Who's responsible for making sure the air you breathe is clean and won't make you sick? Some may say it's the job of governments, corporations, or even scientists. In truth, we are *all* responsible for helping keep our air clean and safe.

We wouldn't be able to live on Earth if the atmosphere was too polluted. However, we can't entirely control the quality of the air we breathe. Natural pollutants get into the atmosphere and affect our breathing. Thick smoke from forest fires can hang in the air and irritate our lungs. Invisible gases from volcanic eruptions can be dangerous as well.

Even though there are many natural air pollutants, humans also cause serious damage to the quality of the air. For centuries, air pollution has been a growing concern around the world. If you can picture a city skyline with factory smoke pumping into the air, you can imagine the challenges of combating air pollution. Growing populations need energy, and they need goods that are produced in factories. However, these processes often pollute the air.

Air pollution is serious and can be deadly. Fortunately, people have made changes over the past century—and especially over the past 50 or 60 years—that have saved countless lives and improved air quality for millions of people around the world. The actions of individuals can make a large impact on the air quality for all people. You can take part in the fight for clean air too!

People should be able to enjoy clean, fresh air even when they're in big cities.

According to the American Lung Association, pollution particles can be smaller than 1/30th the diameter of a human hair. This makes them easy to breathe in through the nose or mouth, and such tiny particles can easily enter the lungs as well.

Breathing in polluted air is linked to many illnesses. It increases the risk of lung cancer, it irritates asthma, and it's linked to some types of heart disease. Polluted air has also been linked to strokes, autism, diabetes, and lower IQ.

A disease known as bronchitis tightens the pathways to the lungs and makes breathing difficult. During a time when Great Britain's air was very polluted at the turn of the 20th century, bronchitis was the leading cause of death there.

Even with improvements in air quality over the past century, air pollution is still a major cause of death around the world. According to the Environmental Defense Fund (EDF), air pollution causes 6.4 million deaths per year, and more than a half million of those people are children. Even with limits on outdoor air pollution, the World Health Organization (WHO) estimates that 300 million children live where air pollution is more than six times the international limit.

Air pollution affects people differently based on age and location. Children and the elderly are affected the most by poor air quality because their lungs are still forming or may be weak. People living in cities or near busy highways are at a greater risk of disease due to air pollution.

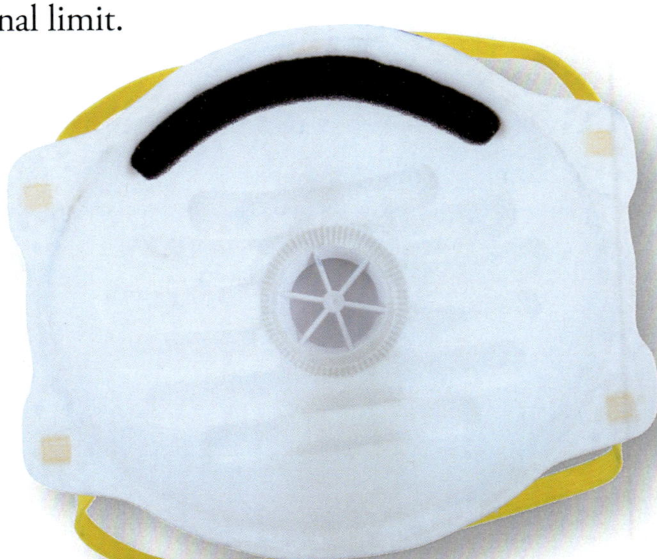

RESPIRATORY MASK

Air pollution and poor air quality damage our lungs and cause disease.

CHAPTER TWO

AN ONGOING PROBLEM

Air quality has been an issue since the times of early civilizations. Historians have found ancient Egyptian mummies with evidence of blackened lungs. Since humans began using fire, smoke and particles from burned materials have created air pollution.

Heat from fires makes the thermal energy needed for many processes, such as cooking and creating raw materials. In time, ancient civilizations began to burn Earth materials to remove metals. This is a process called smelting. During the Middle Ages, smelting to obtain metals such as copper became common throughout Europe. The process filled the air with thick, heavy smoke and soot.

Burning peat is another process that affected air quality in the Middle Ages. Peat is a soil-like material made of partly broken-down plants. It's sometimes the first stage in coal formation. Burning peat produced the high temperatures needed to make glass, bricks, and tiles. Burning peat was popular until the mid-1700s when people began burning coal for energy.

Coal allowed for the start of the Industrial Revolution in the 18th century, when people began using more machines and building factories. Soon factory smokestacks became part of the skylines of large cities. In the United States, factories and mill towns cropped up and poured thick, dangerous smoke into heavily populated areas. Even though the cities of the Industrial Revolution were filled with thick smoke that made people ill, the smog was seen as a sign of prosperity and a booming economy.

Workers at the Solvay Process Company plant, pictured here in Syracuse, New York, during the late 1800s, were at risk for developing lung cancer from asbestos.

To get a clear picture of how widespread the problem of air pollution is in our lives, just think about the impact of cars. We rely on cars to get around, but cars have been polluting our atmosphere for about a century.

Making gasoline involves heating crude oil, a fossil fuel. This oil is a natural resource that takes millions of years to form and must be taken from under the ground. Then the oil must be processed through factory systems that pollute the air and contribute to greenhouse gases in the atmosphere. These gases include carbon dioxide, methane, nitrous oxide, and others. The most abundant of these, however, is carbon dioxide.

The story of cars and their contribution to air pollution doesn't stop with making gasoline. Emissions from cars are harmful as well. According to the U.S. Environmental Protection Agency (EPA), the average passenger car emits nearly 19.6 pounds (8.9 kg) of carbon dioxide, or CO_2, for every gallon of gas burned. Multiply that by the number of cars in the world. Today, there are around 278 million vehicles on roads in the United States alone.

According to the EPA, some greenhouse gases can stay trapped in the atmosphere for thousands of years. These gases then trap heat from the sun and keep it in the atmosphere. Some trapped greenhouse gases keep Earth warm enough to sustain life—but if there are too many, the planet starts to grow too warm.

Emissions from cars contribute to high levels of greenhouse gases in the atmosphere and help create smog over crowded cities.

CHAPTER THREE

A GLOBAL HEALTH ISSUE

As the number of cars on the road grew throughout the 20th century, the number of factories continued to grow as well. Smokestacks spewed thick, dark smoke in a number of places in Europe and the United States. This smoke hung in the air, and some people who breathed it in became ill.

In October 1948, the normally smoggy air over the mill town of Donora, Pennsylvania, became deadly. In a rare weather event, a layer of warm air trapped cold, smoggy air underneath it. It held the smog from the town's 10 smokestacks near ground level. The smog was so thick that residents couldn't see the traffic lights on the street and had trouble seeing the ground when they were walking. Over a four-day period, 20 people died because of breathing issues and about 6,000 others became sick. The local zinc plant that people believe created the issues didn't shut down until several days into the incident—and reopened only a day later.

DONORA, PENNSYLVANIA, 1948

Then, four years after Donora's killer smog, the Great London Smog of 1952 spread across the city in Great Britain. The smog itself lasted five days, but people became so ill that the event caused at least 4,000 deaths over a period of just a few weeks—and maybe as many as 12,000 deaths. Factory emissions remained trapped near ground level and became so thick that ambulances couldn't find their way to help people. People abandoned their cars on the streets. After the smog lifted, the eyes of the world were upon London.

This photograph was taken during the Great London Smog of 1952. It shows children playing in the smog at Hampstead Heath.

London had long been a place with heavily polluted air. Fogs that caused days with low visibility were jokingly referred to as "pea-soupers." The term "London fog" was well known even before the turn of the 20th century. However, environmental disasters such as the ones in London, Great Britain, and Donora, Pennsylvania, finally brought real attention to the issue of industrial smog and its impact on human health.

Investigators studied the disaster in Donora and began monitoring, or keeping an eye on, air quality. They released a detailed report about the problem, and in 1950, President Harry Truman led the first national air pollution conference. In 1963, Congress finally passed the first Clean Air Act. President Lyndon B. Johnson signed it into law.

HARRY TRUMAN

Similar action was taken in Great Britain after the Great London Smog. In 1956, Parliament passed the Clean Air Act, which reduced the amount of coal that industries and individuals could burn.

These laws were the first of many with the goal of helping to improve air quality and save lives. The U.S. Clean Air Act has been amended several times, requiring companies to meet standards for air quality. By 1970, it had policies with the goal of reducing harmful gas emissions by nearly 90 percent.

There was another major amendment to the Clean Air Act in 1990. It addressed issues such as damage to the ozone layer and acid rain. Acid rain is rain that has high levels of acid due to harmful gases in the atmosphere.

Because of the Clean Air Act of 1963, federal and state governments have more control over standards for car emissions.

CHAPTER FOUR

IMPROVEMENTS AROUND THE WORLD

New vehicle and fuel technology has cut vehicle emissions by 90 percent since 1998, and standards in some places are still getting stricter. The United Nations (UN) and WHO have also helped with the improvements. Organizations around the world are seeing the importance of paying attention to industrial and vehicle emissions. People and groups are realizing that it's not enough to measure and study these emissions—we must take action and make improvements.

Public awareness and action have led to the changes so far. In some ways, things have changed since the Donora disaster, when families were afraid to hold the company accountable because many relied on the zinc plant for jobs. Now corporations must follow laws about pollution. They must also be concerned about losing customers or damaging their reputation if they're the cause of an environmental problem. This is good news because it gives citizens more power to create change.

HEADQUARTERS OF THE WORLD HEALTH ORGANIZATION

However, the fight for clean air is far from over. Studies by the UN, WHO, and other international agencies are uncovering new air quality problems in more recently industrialized countries such as China, India, Thailand, Mexico, and Brazil. Over the past several decades, these countries have been producing and shipping goods to growing populations around the world. This adds to the pollution already in the air.

Leaders and activists around the world regularly work together to propose new and lasting changes to air safety regulations.

With the success of air pollution legislation, people began turning their attention to the problem of indoor air quality. Smoke from cigarettes, cigars, and other devices pollutes the air in a similar way to car emissions or factory smoke. Tobacco smoke is harmful not just to smokers but also to anyone breathing in the air nearby.

Pollutants such as asbestos and mold are dangerous as well. We may not even know we're breathing them in. According to the EPA, repeated exposure to these pollutants might give people headaches, dizziness, and fatigue or even cause cancer. They may irritate the nose, throat, or eyes. It could take years for symptoms to appear.

Other indoor air pollutants include the chemicals in many cleaning products and the finishes on new flooring, carpeting, or furniture. When using chemical products indoors, it's important to have as much fresh, clean air available as possible. That may mean keeping windows open or using fans to make sure chemicals don't stay concentrated in the air.

In many commercial buildings, vents that pump air throughout a building can help keep the air moving. However, the air must still be kept safe and clean. Legislation about indoor air quality can help with this.

Laws also prevent people from smoking in many indoor places. Some places have banned harmful substances such as lead, some pesticides, and chemical products indoors.

Water damage can lead to mold growth indoors.

CHAPTER FIVE

LINKS TO CLIMATE CHANGE

Air pollution isn't only damaging to human health: it's also harmful to the environment. Black carbon and methane are two of many air pollutants that put the environment in danger. Methane leads to something called ground-level ozone, which kills plants and causes diseases.

Over its history, Earth has gone through natural periods of climate change. However, according to NASA scientists and others, studies indicate that Earth is now in a period of climate change that is almost definitely the result of human activities. These activities, including industrialization, have damaged air quality and caused dramatically increased levels of greenhouse gases in the atmosphere since the mid-19th century.

In our current period of climate change, temperatures around the world have been steadily rising. According to NASA, Earth's average surface temperature is 1.62°Fahrenheit (0.9°C) higher than it was in the late 19th century. Five of the warmest years on record have occurred since 2010. Other signs of climate change include warming oceans, rising sea levels, and shrinking ice sheets and glaciers. One of the most noticeable and destructive signs of climate change at this point is the increase in extreme weather events, such as hurricanes and flooding from intense rainfall. Reducing emissions and limiting air pollution would play a big part in slowing climate change.

In 2014, many protesters participated in the People's Climate March. They wanted to change the government's environmental policies.

CHAPTER SIX

MAKE THE WORLD TAKE NOTICE

There are many ways to make people aware of the importance of clean air. Protests help make people aware of issues. Boycotts prevent money from going to companies that violate environmental laws. Writing to government leaders can show support for or disagreement over possible laws.

Educating people about the state of the planet is important. One way that people raise awareness of important issues is through art. In September 2019, British artist Michael Pinsky brought his environmental art display to land outside the United Nations Climate Action Summit in New York City. Pinsky's *Pollution Pods* is a five-dome sculpture that viewers walk through to experience air quality conditions in five cities around the world.

The climate-controlled experience simulates the smells and conditions of each city. Visitors start at the pod with the best air quality—one depicting a location on a Norwegian island. As visitors pass through the other pods, they experience some of the conditions of pollution and air quality in four other locations: London, Great Britain; New Delhi, India; Sao Paolo, Brazil; and Beijing, China. Each pod also contains information on how that city's air quality impacts human health.

The project was commissioned by the Norwegian University of Science and Technology and represents results of a four-year study about air quality and its impact. It was designed to get people talking about important environmental issues.

The art installation *Pollution Pods* has traveled around the world. When it was at the UN Climate Action Summit, activists and policymakers spoke with visitors about climate change.

CHAPTER SEVEN

WHAT'S BEING DONE?

In recent years, some laws have forced corporations to react to air pollution and its effects on climate change. New lower-emission power plants and cleaner energy sources are part of these reactions. Instead of burning fossil fuels, more companies are looking for ways to use renewable energy sources such as solar and wind power. These energy sources emit less carbon dioxide and other greenhouse gases into the atmosphere.

Another new process called carbon capture and storage also helps reduce greenhouse gases. During this process, carbon dioxide is trapped as it's released from its source, usually a power plant. Trucks or pipelines then transport the gas to a storage location, such as an underground facility. Storing the carbon dioxide will keep it from entering the atmosphere. This will help fight air pollution and climate change. Many scientists and environmentalists consider carbon capture a promising way to reduce carbon emissions from older factories still burning fossil fuels.

Some companies use new and more environmentally friendly ways to work. They're setting an example for other companies and setting up new processes for the future. New systems such as carbon capture are increasingly important, as new regulations may further limit the pollutants released into the atmosphere.

New technologies and more environmentally friendly methods can be especially helpful in developing countries where air quality doesn't meet standards for healthy living.

Carbon capture technology is one of several methods that may be able to help us deal with air pollution and climate change.

The improvements made to air quality over the past few decades can be partly credited to government action. However, there's a lot more that governments can do. World leaders gather regularly to discuss issues such as air quality, pollution, and climate change. Each nation makes its own decisions about how to act.

As a result, some nations have stricter air quality standards than others. Countries such as China and India are experiencing the economic growth that European countries and the United States experienced during the 19th and 20th centuries. The air quality standards in these countries are less strict than WHO's guidelines. This causes a problem for governments with strict guidelines. We must view air pollution and climate change on a global scale for real and permanent change to be made.

Even within the United States, there are differences in air quality. According to the American Lung Association, almost half of the counties in the United States still have air quality below the target standard. The organization's "State of the Air 2019" study found that some areas of the United States had higher levels of short-term pollution in recent years than they had ever had before. Short-term air pollution may include that caused by wildfires, chemical spills, or other incidents. While there have been improvements in some forms of long-term pollution, short-term air pollution still needs to be further studied and addressed.

Activists and world leaders met at the United Nations Climate Action Summit in New York City during Climate Week in 2019.

Perhaps the most important way to make changes in air quality is to become an advocate for the environment. When people get involved in making change, governments and corporations take notice. Individual citizens elect politicians and purchase the products made by corporations. Individuals have the power to make a difference through advocacy.

Young people can play a part in creating change too. The news media has taken notice of Swedish teen Greta Thunberg and her advocacy about climate change. She is one of many young people who feel that the adults in power aren't doing enough to address environmental issues. In 2019, Thunberg was named *Time* magazine's Person of the Year, making her the youngest person to ever win the title.

Some young people have started advocacy groups that can bring people together for a common cause. At 16, American student Isra Hirsi helped establish the U.S. Youth Climate Strike. Hirsi and other teens were inspired by news they heard about the environment. The work of these young activists prompts others to act as well.

The Youth Climate Strike group organizes strikes around the country to bring attention to climate change. They support the Green New Deal, which is proposed legislation to address climate change. Among other things, it could make it possible for people impacted by disasters related to climate change to apply for government aid and support.

ISRA HIRSI

Young activists such as Greta Thunberg and Isra Hirsi are making a difference by uniting people and encouraging action to help the environment.

CHAPTER EIGHT

A BREATH OF FRESH AIR

We're already decades into the fight to improve Earth's air quality after years of damage. While there have been many successes, more work still needs to be done. The more people from around the world can come together to encourage change, the louder their voices will be.

As some governments act to legislate reductions in greenhouse gas emissions and dangerous particle pollution, corporations will have to improve their activities and reduce air pollution. As young activists have asserted during protests and strikes, the future of the planet depends on people acting today. Whether people fight for change through protest, education, or even through art, the most important thing is to act.

Environmental disasters are going to continue, but there are other worries as well. In 2020, the COVID-19 pandemic struck the world and changed it forever. COVID-19 cases were worse and more common in areas where air pollution was particularly bad. At the same time, the reduction in economic activity during this time resulted in a decrease in air pollution in some areas.

New technologies, public awareness, and strong legislation are the best ways to improve air quality in the future. Using clean energy solutions instead of burning fossil fuels is the best way to ensure factory and car emissions are reduced or eliminated.

In the future, there may be a brighter outlook for our country and our world because of work that people are doing today. The young people of today will be the government, corporate, and environmental leaders of tomorrow.

GLOSSARY

advocate (AD-vuh-kuht) A person who argues for or supports a cause or policy.

asbestos (az-BEH-stus) A natural material that doesn't burn and was once used in building but can harm people when breathed in.

asthma (AZ-muh) A condition that makes it hard for a person to breathe.

autism (AW-tiz-uhm) A developmental disorder that can affect social interactions, thoughts, and behaviors.

bronchitis (brahn-KY-tis) A medical condition caused by swelling within the air passages of the lungs.

diabetes (DY-uh-bee-teez) A disease affecting the body's ability to produce or respond to the hormone insulin.

diameter (dy-AA-muh-tur) The measurement across the center of a round object.

disaster (dih-ZAS-tuhr) Something that happens suddenly and causes much suffering and loss for many people.

fossil fuel (FAH-suhl FYOOL) A fuel—such as coal, oil, or natural gas—that is formed in the earth from dead plants or animals.

greenhouse gas (GREEN-howz GASS) A gas in the atmosphere that traps energy from the sun.

Industrial Revolution (in-DUH-stree-uhl reh-vuh-LOO-shuhn) An era of social and economic change marked by advances in technology and science.

irritate (IHR-uh-tayt) To bother or trouble.

peat (PEET) A soil-like material made of partly broken-down vegetable matter, sometimes the first stage in the formation of coal.

smelt (SMEHLT) To melt rock that contains metal to get the metal out.

smog (SMAWG) Fog mixed with smoke.

stroke (STROHK) A sudden blockage or break of a blood vessel in the brain.

technology (tek-NAH-luh-jee) A method that uses science to solve problems and the tools used to solve those problems.

thermal (THUHR-muhl) Having to do with heat.

INDEX

PRIMARY SOURCE LIST

Page 9
Solvay Process Company. Photograph. Created by Detroit Publishing Co. Published in 1901. Syracuse, New York. Now kept at the Library of Congress Prints and Photographs Division Washington, D.C.

Page 13
The Great Smog of London. Photograph. December 7, 1952. London, England.

Page 15
President Lyndon B. Johnson signing the 1967 Clean Air Act in the East Room of the White House. Photograph. November 21, 1967. Washington, D.C.

WEBSITES

Due to the changing nature of Internet links, Rosen Publishing has developed an online list of websites related to the subject of this book. This site is updated regularly. Please use this link to access the list: www.powerkidslinks.com/SOGI/cleanair